Subtract to solve each problem. Cross out the picture the answers.

a. 10 – 0 =

b. 10 – 1 =

c. 10 – 2 =

d. 10 – 3 =

e. 10 – 4 =

f. 10 – 5 =

Subtract to solve each problem. Cross out the pictures to help you find the answers.

a. 10 − 6 =

b. 10 − 7 =

c. 10 − 8 =

d. 10 − 9 =

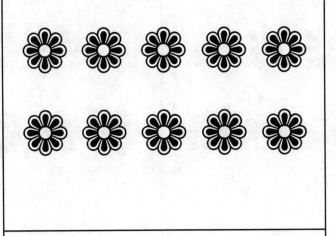

e. 10 − 10 =

Subtract to solve each problem. Cross out the pictures to help you find the answers.

■ ■ ■ ■ ■ ■ ■ ■ ■	■ ■ ■ ■ ■ ■ ■ ■ ■
a. $9 - 0 =$	b. $9 - 1 =$
■ ■ ■ ■ ■ ■ ■ ■ ■	■ ■ ■ ■ ■ ■ ■ ■ ■
c. $9 - 2 =$	d. $9 - 3 =$
■ ■ ■ ■ ■ ■ ■ ■ ■	■ ■ ■ ■ ■ ■ ■ ■ ■
e. $9 - 4 =$	f. $9 - 5 =$
■ ■ ■ ■ ■ ■ ■ ■ ■	■ ■ ■ ■ ■ ■ ■ ■ ■
g. $9 - 6 =$	h. $9 - 7 =$
■ ■ ■ ■ ■ ■ ■ ■ ■	■ ■ ■ ■ ■ ■ ■ ■ ■
i. $9 - 8 =$	j. $9 - 9 =$

Subtract to solve each problem. Cross out the pictures to help you find the answers.

★ ★ ★ ★ ★ ★ ★ ★	★ ★ ★ ★ ★ ★ ★ ★	★ ★ ★ ★ ★ ★ ★ ★
a. 8 − 0 =	b. 8 − 1 =	c. 8 − 2 =
★ ★ ★ ★ ★ ★ ★ ★	★ ★ ★ ★ ★ ★ ★ ★	★ ★ ★ ★ ★ ★ ★ ★
d. 8 − 3 =	e. 8 − 4 =	f. 8 − 5 =
★ ★ ★ ★ ★ ★ ★ ★ ★ ★	★ ★ ★ ★ ★ ★ ★ ★ ★ ★	★ ★ ★ ★ ★ ★ ★ ★ ★ ★
g. 8 − 6 =	h. 8 − 7 =	i. 8 − 8 =

Subtract to solve each problem. Cross out the pictures to help you find the answers.

● ● ● ● ● ● ●	● ● ● ● ● ● ●	● ● ● ● ● ● ●
a. 7 – 0 =	b. 7 – 1 =	c. 7 – 2 =
● ● ● ● ● ● ●	● ● ● ● ● ● ●	● ● ● ● ● ● ●
d. 7 – 3 =	e. 7 – 4 =	f. 7 – 5 =
● ● ● ● ● ● ●	● ● ● ● ● ● ●	▲ ▲ ▲ ▲ ▲ ▲
g. 7 – 6 =	h. 7 – 7 =	i. 6 – 0 =
▲ ▲ ▲ ▲ ▲ ▲	▲ ▲ ▲ ▲ ▲ ▲	▲ ▲ ▲ ▲ ▲ ▲
j. 6 – 1 =	k. 6 – 2 =	l. 6 – 3 =
▲ ▲ ▲ ▲ ▲ ▲	▲ ▲ ▲ ▲ ▲ ▲	▲ ▲ ▲ ▲ ▲ ▲
m. 6 – 4 =	n. 6 – 5 =	o. 6 – 6 =

Subtract to solve each problem. Cross out the pictures to help you find the answers.

�֍ �֍ �֍ �֍ ✖	 ✖ ✖ ✖ ✖ ✖
a. 5 – 0 =	b. 5 – 1 =
 ✖ ✖ ✖ ✖ ✖	✖ ✖ ✖ ✖ ✖
c. 5 – 2 =	d. 5 – 3 =
✖ ✖ ✖ ✖ ✖	✖ ✖ ✖ ✖ ✖
e. 5 + 4 =	f. 5 – 5 =
 ▼ ▼ ▼ ▼	▼ ▼ ▼ ▼
g. 4 – 0 =	h. 4 – 1 =
 ▼ ▼ ▼ ▼	 ▼ ▼ ▼ ▼
i. 4 – 2 =	j. 4 – 3 =

k. 4 – 4 =

Subtract to solve each problem. Cross out the pictures to help you find the answers.

a. 3 – 0 =	b. 3 – 1 =

c. 3 – 2 =	d. 3 – 3 =
e. 2 – 0 =	f. 2 – 1 =
g. 2 – 2 =	h. 2 – 3 =
i. 1 – 1 =	j. 0 – 0 =

a. Cross out 4 sandwiches.

How many are left?_____

b. Cross out 2 peanuts.

How many are left?_____

c. Cross out 1 pineapple.

How many are left?_____

d. Cross out 5 worms.

How many are left?_____

e. Cross out 3 shells.

How many are left?_____

f. Cross out 0 clocks.

How many are left?_____

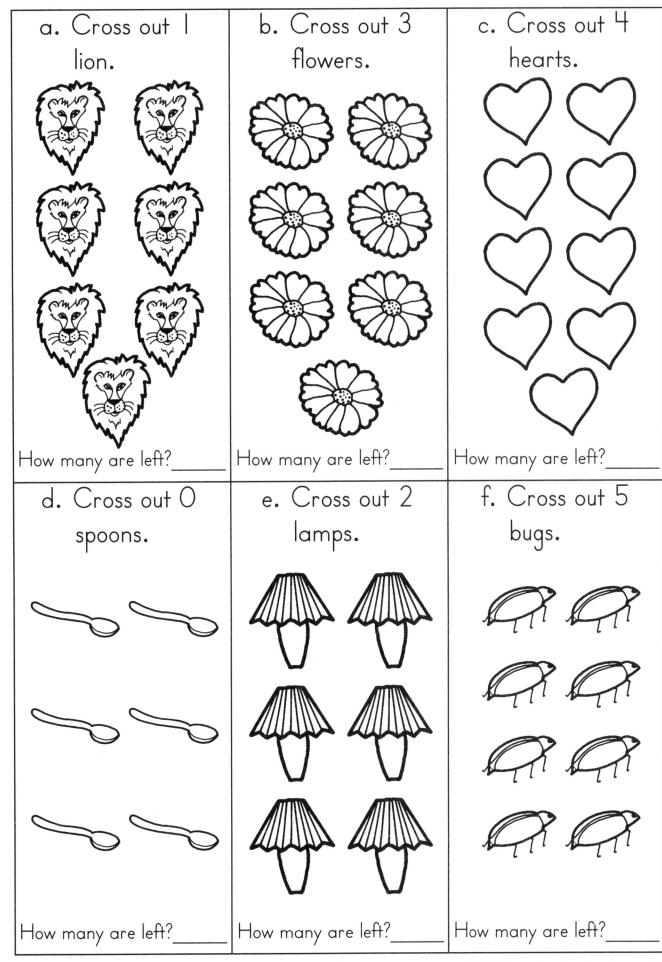

a. Cross out 1 lion.

How many are left?_____

b. Cross out 3 flowers.

How many are left?_____

c. Cross out 4 hearts.

How many are left?_____

d. Cross out 0 spoons.

How many are left?_____

e. Cross out 2 lamps.

How many are left?_____

f. Cross out 5 bugs.

How many are left?_____

a. Count the balloons. Cross out 3 balloons. Write the numerals on the lines to show the subtraction problem and answer.

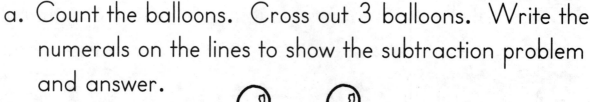 _____ – _____ = _____

b. Count the monkeys. Cross out 5 monkeys. Write the numerals on the lines to show the subtraction problem and answer.

 _____ – _____ = _____

c. Count the mice. Cross out 2 mice. Write the numerals on the lines to show the subtraction problem and answer.

 _____ – _____ = _____

d. Count the bees. Cross out 4 bees. Write the numerals on the lines to show the subtraction problem and answer.

 _____ – _____ = _____

e. Count the hands. Cross out 0 hands. Write the numerals on the lines to show the subtraction problem and answer.

 _____ – _____ = _____

a. Count the cars. Cross out 2 cars. Write the numerals on the lines to show the subtraction problem and answer.

_____ – _____ = _____

b. Count the bears. Cross out 4 bears. Write the numerals on the lines to show the subtraction problem and answer.

_____ – _____ = _____

c. Count the books. Cross out 3 books. Write the numerals on the lines to show the subtraction problem and answer.

 _____ – _____ = _____

d. Count the fish. Cross out 1 fish. Write the numerals on the lines to show the subtraction problem and answer.

_____ – _____ = _____

e. Count the rings. Cross out 2 rings. Write the numerals on the lines to show the subtraction problem and answer.

_____ – _____ = _____

a. Count the apples. Cross out 4 apples. Write the numerals on the lines to show the subtraction problem and answer.

_____ − _____ = _____

b. Count the baskets. Cross out 2 baskets. Write the numerals on the lines to show the subtraction problem and answer.

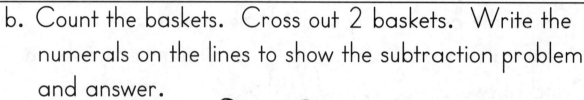

_____ − _____ = _____

c. Count the pens. Cross out 3 pens. Write the numerals on the lines to show the subtraction problem and answer.

_____ − _____ = _____

d. Count the chicks. Cross out 4 chicks. Write the numerals on the lines to show the subtraction problem and answer.

_____ − _____ = _____

e. Count the snakes. Cross out 1 snake. Write the numerals on the lines to show the subtraction problem and answer.

_____ − _____ = _____

Cross out each answer in the hamburger as you solve the problems.

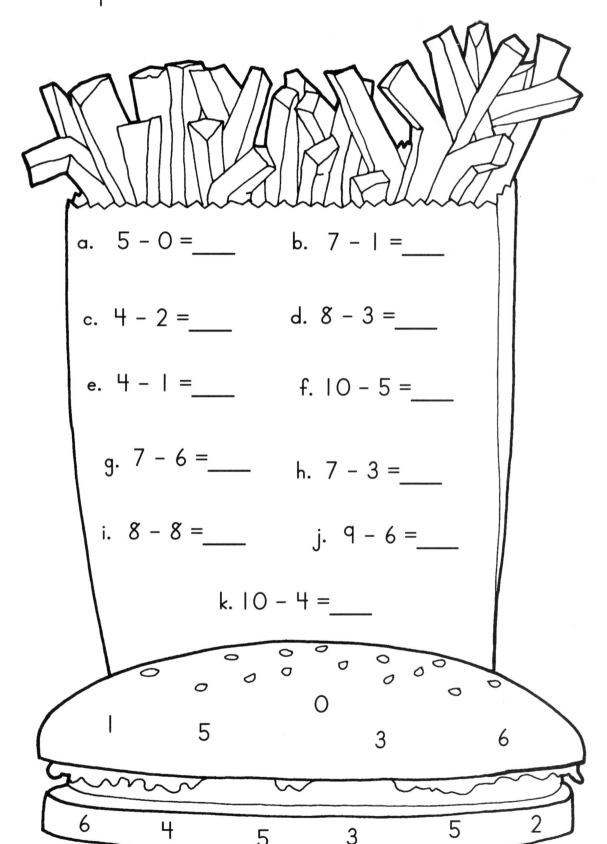

a. 5 – 0 = ___

b. 7 – 1 = ___

c. 4 – 2 = ___

d. 8 – 3 = ___

e. 4 – 1 = ___

f. 10 – 5 = ___

g. 7 – 6 = ___

h. 7 – 3 = ___

i. 8 – 8 = ___

j. 9 – 6 = ___

k. 10 – 4 = ___

1   5   0   3   6

6   4   5   3   5   2

Cross out each answer in the tool box as you solve the problems.

a. 6 − 0 = ___

b. 10 − 1 = ___

c. 3 − 2 = ___

d. 5 − 3 = ___

e. 7 − 4 = ___

f. 9 − 5 = ___

g. 8 − 6 = ___

h. 7 − 2 = ___

i. 8 − 2 = ___

j. 9 − 5 = ___

k. 10 − 9 = ___

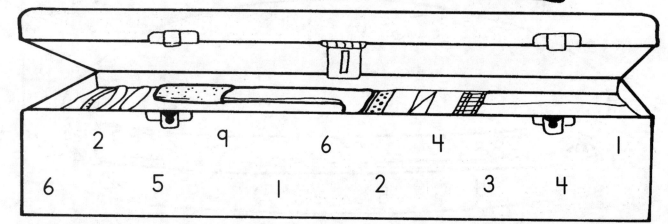

2   9   6   4   1
6   5   1   2   3   4

Cross out each answer in the crayon box as you solve the problems.

a.

1 – 0 = ____

b.

8 – 1 = ____

c.

5 – 2 = ____

d.

4 – 3 = ____

e.

8 – 4 = ____

f.

5 – 4 = ____

g.

9 – 6 = ____

h.

7 – 7 = ____

i.

8 – 7 = ____

j.

10 – 5 = ____

k.

9 – 3 = ____

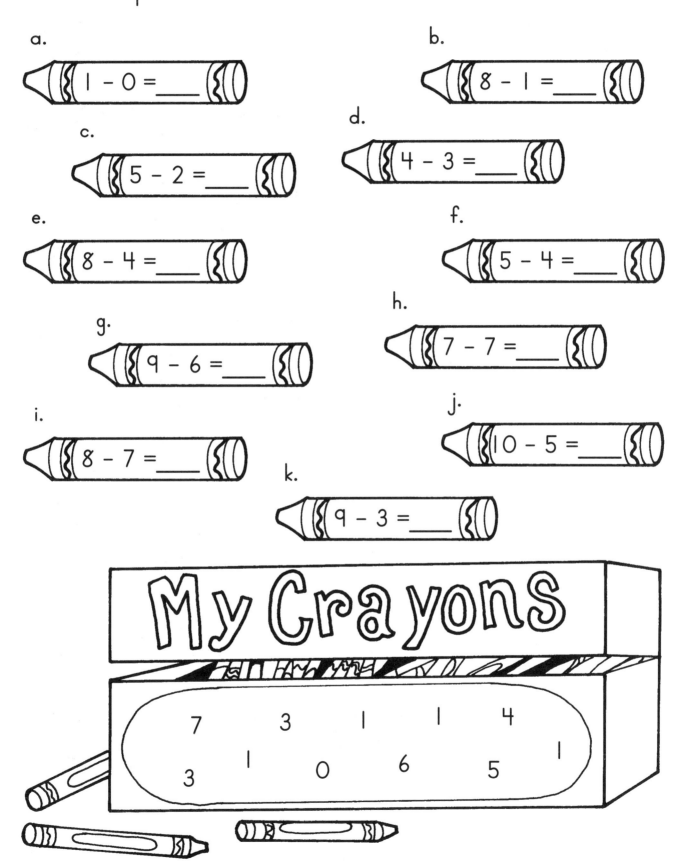

My Crayons

| 7 | 3 | 1 | 1 | 4 |
| 3 | 1 | 0 | 6 | 5 | 1 |

Draw a line to match each equation to its answer.

a. 9 – 2 = _____

b. 7 – 4 = _____

c. 10 – 9 = _____

d. 6 – 6 = _____

e. 3 – 1 = _____

f. 8 – 4 = _____

Draw a line to match each equation to its answer.

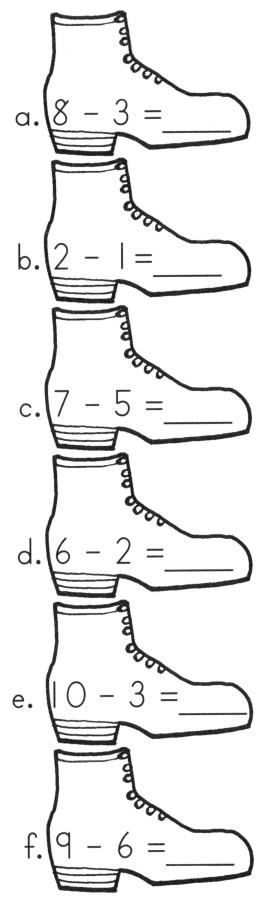

a. 8 – 3 = _____

b. 2 – 1 = _____

c. 7 – 5 = _____

d. 6 – 2 = _____

e. 10 – 3 = _____

f. 9 – 6 = _____

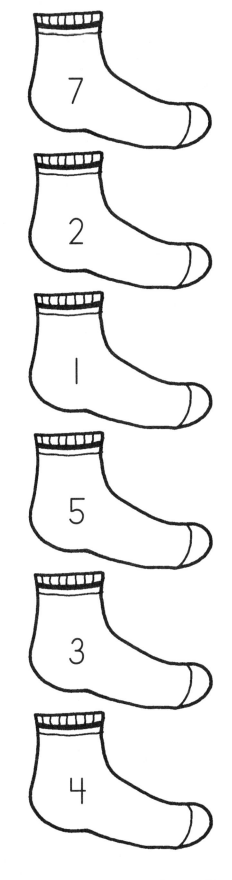

7

2

1

5

3

4

#2253 Subtraction Practice Facts to 10

Use the pictures to solve these problems.

a.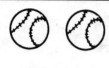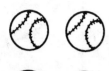

$$6 - 2 = \underline{\qquad}$$
_____ - _____ = _____

b.

_____ - _____ = _____

c.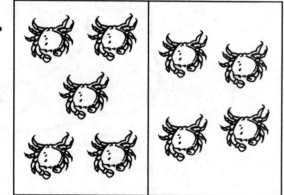

_____ - _____ = _____

d.

_____ - _____ = _____

e.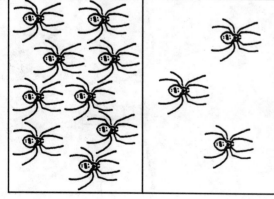

_____ - _____ = _____

f.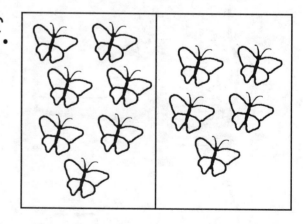

_____ - _____ = _____

Use the pictures to solve these problems.

a.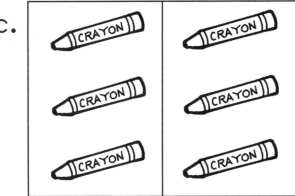

$$\underline{\quad 4 \quad} - \underline{\quad 3 \quad} = \underline{\qquad}$$

b.

$$\underline{\qquad} - \underline{\qquad} = \underline{\qquad}$$

c.

$$\underline{\qquad} - \underline{\qquad} = \underline{\qquad}$$

d.

$$\underline{\qquad} - \underline{\qquad} = \underline{\qquad}$$

e.

$$\underline{\qquad} - \underline{\qquad} = \underline{\qquad}$$

f.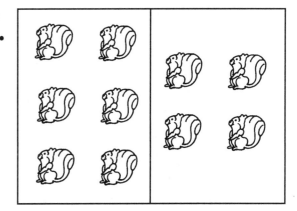

$$\underline{\qquad} - \underline{\qquad} = \underline{\qquad}$$

Use the pictures to solve these problems.

a.

$$\underline{}2\underline{}-\underline{}2\underline{}=\underline{}$$

b.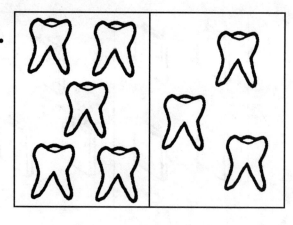

$$\underline{}-\underline{}=\underline{}$$

c.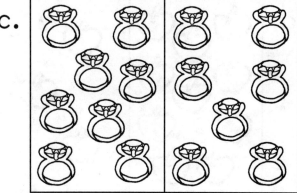

$$\underline{}-\underline{}=\underline{}$$

d.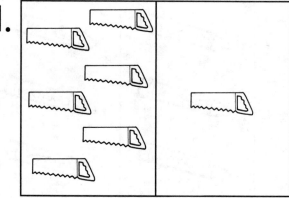

$$\underline{}-\underline{}=\underline{}$$

e.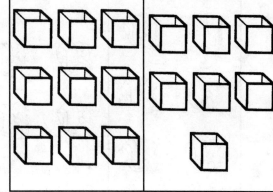

$$\underline{}-\underline{}=\underline{}$$

f.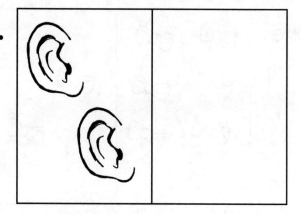

$$\underline{}-\underline{}=\underline{}$$

Help the lion tamer find his lion. Solve the problems. Draw a line from the tamer to the lion with the answer of 3. Color the lion tamer.

Help the train find its caboose. Solve the problems. Draw a line from the engine to the caboose with the answer of 4. Color the train.

$$\begin{array}{r} 6 \\ -3 \\ \hline \end{array}$$

$$\begin{array}{r} 9 \\ -1 \\ \hline \end{array}$$

$$\begin{array}{r} 10 \\ -4 \\ \hline \end{array}$$

4

$$\begin{array}{r} 8 \\ -4 \\ \hline \end{array}$$

$$\begin{array}{r} 8 \\ -3 \\ \hline \end{array}$$

$$\begin{array}{r} 7 \\ -5 \\ \hline \end{array}$$

Help the girl find her kite. Solve the problems.
Draw a line from the girl to the kite with the
answer of 6. Color the girl.

Guess what is in the box. Find the answers. Then write the letter in each box that matches each answer. Read the word you spell and draw it in the box.

1	2	3	4	5
n	p	a	e	l

6	5		4	6	5	10	9
−3	−4		−1	−4	−3	−5	−5
3							
a							

Guess what is in the box. Find the answers. Then write the letter in each box that matches each answer. Read the word you spell and draw it in the box.

2	3	4	5
p	u	a	y

$$\begin{array}{c}7\\-3\\\hline 4\end{array} \qquad \begin{array}{c}6\\-4\\\hline\end{array} \qquad \begin{array}{c}7\\-4\\\hline\end{array} \qquad \begin{array}{c}9\\-7\\\hline\end{array} \qquad \begin{array}{c}3\\-1\\\hline\end{array} \qquad \begin{array}{c}8\\-3\\\hline\end{array}$$

a ☐ ☐ ☐ ☐ ☐

Guess what is in the box. Find the answers. Then write the letter in each box that matches each answer. Read the word you spell and draw it in the box.

0	1	2	3	4	5
l	f	a	o	b	t

$$\begin{array}{c} 7 \\ -5 \\ \hline 2 \end{array}$$

a

$$\begin{array}{c} 8 \\ -7 \\ \hline \end{array}$$

$$\begin{array}{c} 9 \\ -6 \\ \hline \end{array}$$

$$\begin{array}{c} 3 \\ -0 \\ \hline \end{array}$$

$$\begin{array}{c} 6 \\ -1 \\ \hline \end{array}$$

$$\begin{array}{c} 9 \\ -5 \\ \hline \end{array}$$

$$\begin{array}{c} 9 \\ -7 \\ \hline \end{array}$$

$$\begin{array}{c} 1 \\ -1 \\ \hline \end{array}$$

$$\begin{array}{c} 5 \\ -5 \\ \hline \end{array}$$

Find the differences. Color the picture.

5 − 1 = ___

6 − 2 = ___

9 − 3 = ___

10 − 7 = ___

5 − 0 = ___

3 − 1 = ___

7 − 1 = ___

9 − 6 = ___

8 − 2 = ___

6 − 4 = ___

     #2253 Subtraction Practice Facts to 10

Find the answers. Color the pictures.

0 = yellow	1 = pink	2 = brown	3 = green	4 = white

$$\begin{array}{r} 5 \\ -4 \\ \hline \end{array}$$

$$\begin{array}{r} 6 \\ -3 \\ \hline \end{array}$$

$$\begin{array}{r} 4 \\ -3 \\ \hline \end{array}$$

$$\begin{array}{r} 6 \\ -6 \\ \hline \end{array}$$

$$\begin{array}{r} 7 \\ -5 \\ \hline \end{array}$$

$$\begin{array}{r} 7 \\ -3 \\ \hline \end{array}$$

$$\begin{array}{r} 8 \\ -4 \\ \hline \end{array}$$

$$\begin{array}{r} 9 \\ -8 \\ \hline \end{array}$$

$$\begin{array}{r} 5 \\ -3 \\ \hline \end{array}$$

$$\begin{array}{r} 10 \\ -6 \\ \hline \end{array}$$

Read each word problem. Write the number sentence it shows. Find the difference.

a

A dog was walking through his yard when he saw 7 cats on the fence. He barked, and 6 cats ran away. How many cats were left?

$$4 - 2 = 1$$

b

A woman grew a vegetable garden, and in it there were 9 ears of corn. She picked 3 ears on one day, 0 ears on the next, and 6 ears on the third day. How many ears of corn were left in her garden?

c

The monkey had 10 bananas. He ate 1 the first day and 6 the second. How many bananas were left?

d

A boy had 5 assignments for homework. He completed 3 before dinner and 1 after dinner. How many assignments were left?

Find the answers.	j. $2 - 0 =$ ___	t. $9 - 4 =$ ___	dd. $4 - 0 =$ ___
a. $9 - 8 =$ ___	k. $7 - 5 =$ ___	u. $10 - 4 =$ ___	ee. $1 - 0 =$ ___
b. $9 - 7 =$ ___	l. $6 - 0 =$ ___	v. $7 - 1 =$ ___	ff. $4 - 4 =$ ___
c. $10 - 3 =$ ___	m. $7 - 6 =$ ___	w. $7 - 2 =$ ___	gg. $9 - 6 =$ ___
d. $6 - 6 =$ ___	n. $8 - 1 =$ ___	x. $5 - 5 =$ ___	hh. $10 - 0 =$ ___
e. $3 - 0 =$ ___	o. $5 - 4 =$ ___	y. $6 - 5 =$ ___	ii. $8 - 5 =$ ___
f. $8 - 2 =$ ___	p. $9 - 3 =$ ___	z. $8 - 6 =$ ___	jj. $4 - 1 =$ ___
g. $10 - 7 =$ ___	q. $10 - 6 =$ ___	aa. $9 - 1 =$ ___	kk. $2 - 2 =$ ___
h. $10 - 9 =$ ___	r. $6 - 2 =$ ___	bb. $9 - 2 =$ ___	ll. $8 - 7 =$ ___
i. $5 - 1 =$ ___	s. $9 - 0 =$ ___	cc. $10 - 10 =$ ___	mm. $10 - 8 =$ ___

Answer Key

Page 1
a. 10
b. 9
c. 8
d. 7
e. 6
f. 5

Page 2
a. 4
b. 3
c. 2
d. 1
e. 0

Page 3
a. 9
b. 8
c. 7
d. 6
e. 5
f. 4
g. 3
h. 2
i. 1
j. 0

Page 4
a. 8
b. 7
c. 6
d. 5
e. 4
f. 3
g. 2
h. 1
i. 0

Page 5
a. 7
b. 6
c. 5
d. 4
e. 3

f. 2
g. 1
h. 0
i. 6
j. 5
k. 4
l. 3
m. 2
n. 1
o. 0

Page 6
a. 5
b. 4
c. 3
d. 2
e. 1
f. 0
g. 4
h. 3
i. 2
j. 1
k. 0

Page 7
a. 3
b. 2
c. 1
d. 0
e. 2
f. 1
g. 0
h. 1
i. 0
j. 0

Page 8
a. 3
b. 4
c. 3
d. 1
e. 4
f. 5

Page 9
a. 6
b. 4
c. 5
d. 6
e. 4
f. 3

Page 10
a. $6 - 3 = 3$
b. $7 - 5 = 2$
c. $9 - 2 = 7$
d. $10 - 4 = 6$
e. $5 - 0 = 5$

Page 11
a. $6 - 2 = 4$
b. $5 - 4 = 1$
c. $7 - 3 = 4$
d. $9 - 1 = 8$
e. $3 - 2 = 1$

Page 12
a. $9 - 4 = 5$
b. $8 - 2 = 6$
c. $7 - 3 = 4$
d. $8 - 4 = 4$
e. $8 - 1 = 7$

Page 13
a. $5 - 0 = 5$
b. $7 - 1 = 6$
c. $4 - 2 = 2$
d. $8 - 3 = 5$
e. $4 - 1 = 3$
f. $10 - 5 = 5$
g. $7 - 6 = 1$
h. $7 - 3 = 4$
i. $8 - 8 = 0$
j. $9 - 6 = 3$
k. $10 - 4 = 6$

Page 14
a. $6 - 0 = 6$
b. $10 - 1 = 9$
c. $3 - 2 = 1$
d. $5 - 3 = 2$
e. $7 - 4 = 3$
f. $9 - 5 = 4$
g. $8 - 6 = 2$
h. $7 - 2 = 5$
i. $8 - 2 = 6$
j. $9 - 5 = 4$
k. $10 - 9 = 1$

Page 15
a. $1 - 0 = 1$
b. $8 - 1 = 7$
c. $5 - 2 = 3$
d. $4 - 3 = 1$
e. $8 - 4 = 4$
f. $5 - 4 = 1$
g. $9 - 6 = 3$
h. $7 - 7 = 0$
i. $8 - 7 = 1$
j. $10 - 5 = 5$
k. $9 - 3 = 6$

Page 16
a. $9 - 2 = 7$
b. $7 - 4 = 3$
c. $10 - 9 = 1$
d. $6 - 6 = 0$
e. $3 - 1 = 2$
f. $8 - 4 = 4$

Page 17
a. $8 - 3 = 5$
b. $2 - 1 = 1$
c. $7 - 5 = 2$
d. $6 - 2 = 4$
e. $10 - 3 = 7$
f. $9 - 6 = 3$

Answer Key (cont.)

Page 18
a. 6 − 2 = 4
b. 8 − 4 = 4
c. 5 − 4 = 1
d. 2 − 1 = 1
e. 9 − 3 = 6
f. 7 − 5 = 2

Page 19
a. 4 − 3 = 1
b. 9 − 5 = 4
c. 3 − 3 = 0
d. 10 − 3 = 7
e. 7 − 2 = 5
f. 6 − 4 = 2

Page 20
a. 2 − 2 = 0
b. 5 − 3 = 2
c. 8 − 7 = 1
d. 6 − 1 = 5
e. 9 − 7 = 2
f. 2 − 0 = 2

Page 21
3 − 2 = 1
6 − 5 = 1
7 − 2 = 5
4 − 4 = 0
5 − 2 = 3
6 − 4 = 2

Page 22
6 − 3 = 3
9 − 1 = 8
10 − 4 = 6
8 − 4 = 4
8 − 3 = 5
7 − 5 = 2

Page 23
7 − 4 = 3
7 − 1 = 6
8 − 4 = 4
9 − 0 = 9
5 − 3 = 2
6 − 5 = 1

Page 24
6 − 3 = 3 (a)
5 − 4 = 1 (n)
4 − 1 = 3 (a)
6 − 4 = 2 (p)
5 − 3 = 2 (p)
10 − 5 = 5 (l)
9 − 5 = 4 (e)

an apple

Page 25
7 − 3 = 4 (a)
6 − 4 = 2 (p)
7 − 4 = 3 (u)
9 − 7 = 2 (p)
3 − 1 = 2 (p)
8 − 3 = 5 (y)

a puppy

Page 26
7 − 5 = 2 (a)
8 − 7 = 1 (f)
9 − 6 = 3 (o)
3 − 0 = 3 (o)
6 − 1 = 5 (t)
9 − 5 = 4 (b)
9 − 7 = 2 (a)
1 − 1 = 0 (l)
5 − 5 = 0 (l)

a football

Page 27
5 − 1 = 4 (brown)
6 − 2 = 4 (brown)
9 − 3 = 6 (yellow)
10 − 7 = 3 (orange)
5 − 0 = 5 (blue)
3 − 1 = 2 (green)
7 − 1 = 6 (yellow)
9 − 6 = 3 (orange)
8 − 2 = 6 (yellow)
6 − 4 = 2 (green)

Page 28
5 − 4 = 1 (pink)
6 − 3 = 3 (green)
4 − 3 = 1 (pink)
6 − 6 = 0 (yellow)
7 − 5 = 2 (brown)
7 − 3 = 4 (white)
8 − 4 = 4 (white)
9 − 8 = 1 (pink)
5 − 3 = 2 (brown)
10 − 6 = 4 (white)

Page 29
a. 7 − 6 = 1
b. 9 − 3 − 0 − 6 = 0
c. 10 − 1 − 6 = 3
d. 5 − 3 − 1 = 1

Page 30
a. 9 − 8 = 1
b. 9 − 7 = 2
c. 10 − 3 = 7
d. 6 − 6 = 0
e. 3 − 0 = 3
f. 8 − 2 = 6
g. 10 − 7 = 3
h. 10 − 9 = 1
i. 5 − 1 = 4
j. 2 − 0 = 2
k. 7 − 5 = 2
l. 6 − 0 = 6
m. 7 − 6 = 1
n. 8 − 1 = 7
o. 5 − 4 = 1
p. 9 − 3 = 6
q. 10 − 6 = 4
r. 6 − 2 = 4
s. 9 − 0 = 9
t. 9 − 4 = 5
u. 10 − 4 = 6
v. 7 − 1 = 6
w. 7 − 2 = 5
x. 5 − 5 = 0
y. 6 − 5 = 1
z. 8 − 6 = 2
aa. 9 − 1 = 8
bb. 9 − 2 = 7
cc. 10 − 10 = 0
dd. 4 − 0 = 4
ee. 1 − 0 = 1
ff. 4 − 4 = 0
gg. 9 − 6 = 3
hh. 10 − 0 = 10
ii. 8 − 5 = 3
jj. 4 − 1 = 3
kk. 2 − 2 = 0
ll. 8 − 7 = 1
mm. 10 − 8 = 2